MEL BAY PRESENTS

CHORD MELODY METHOD FOR ACCORDION

and Other Keyboard Instruments

By Gary Dahl

CD Contents			
1 Introduction	[1:37]	9 Page 11	[1:09]
2 Page 4	[1:50]	10 Diminished Chords	[5:31]
3 Page 5	[5:40]	11 Page 12	[0:52]
4 Page 6	[13:05]	12 Bye Bye Blues	[12:09]
5 Page 7	[6:29]	13 Page 13	[10:28]
6 Page 8	[3:58]	14 Page 15	[2:51]
7 Page 9	[3:05]	15 Hindustan	[2:13]
8 Page 10	[1:07]	16 Page 16	[6:46]

1 2 3 4 5 6 7 8 9 0

Visit us on the Web at www.melbay.com — E-mail us at email@melbay.com

Table of Contents

Foreword 3
Why We Must Study Harmony 4
Triads 5
First Lead Sheet Etude 6
C Major and A Minor Etudes 7
F Major, Bb Major and D Minor Etudes 8
Song Of The Islands 9
Song Of The Volga Boatmen 9
Augmented Triads with Etudes 10
Diminished Chords 11
Bye Bye Blues 12
Lies 13
Melody Of Love 13
Drifting and Dreaming 14
Linger Awhile 15
Hindustan 15
The Major and Minor 6th Four Part Chords 16
Bye Bye Blackbird 16
The Minor 7th Chord with Etudes 17
Peg O' My Heart with Bass Clef 18
Swanee with Bass Clef 19
The Major 7th Chord with Etudes 20
More Dominant 9th Chord Practice .21
Raised and Flatted Ninths 22
Dominant 11th, Augmented 11th Chords and Tri-Tone Subs 23
Open Position Chords with Etudes 24
Passing Chords 25

PRACTICE SONGS WITH EXTENSIVE EXPLANATIONS AND EXAMPLES

You Made Me Love You 26
April Showers 27
What A Friend 28
Stumbling 29
Sweetheart Of Sigma Chi 30
Fascination 31
Chinatown 32
Ja-Da 33
After You've Gone 34
Do It Again 35
My Melancholy Baby 36
I'm Just Wild About Harry 37
Indian Summer 38
Chord Study and Review 40
I'm Forever Blowing Bubbles 41
Georgia 42
A Good Man Is Hard To Find 43
'Way Down Yonder In New Orleans 44
The Japanese Sandman 45
Bill Bailey 46
Ma (He's Making Eyes at Me) 47
Margie 48
Avalon 49
Alexander's Ragtime Band 51
Indiana 52
Reharmonized Indiana 53
Poor Butterfly 54
About the Author 56

Foreword

The Chord Melody Method for Accordion and Other Keyboard Instruments is a breakthrough course that will teach you to quickly learn all of the chords used in today's music and to apply their harmonic applications professionally to any lead sheet. Imagine possessing the ability to develop and compose a professional-sounding arrangement on the spot while improvising confidently as needed! The average dedicated student will become proficient in 6 to 12 months and will experience a dramatic improvement in all areas of musicianship. The Chord Melody Method will enable you to achieve your ultimate music making goals.

Gary Dahl

Harmony: The combining of notes simultaneously to produce chords, and successively, to produce chord progressions. The term is used descriptively to denote notes and chords to be combined, and also prescriptively to denote a system of structural principles governing their combinations. In the latter sense, harmony has its own body of theoretical literature.

The Chord: The harmonic theory of recent times, which evolved gradually between the 16th and 18th centuries, is based on the idea that a chord (three or four different notes sounded simultaneously) is to be taken as a primary, indivisible unit.

Block Chords: (chordal style) - A style in which all parts move in the same rhythm, thus producing a succession of chords.

WHY WE MUST STUDY HARMONY
by Gary Dahl

Music is made up of three elements: rhythm, melody and harmony. Probably only one person out of a thousand does not have a sense of rhythm. One person out of a hundred does not have a sense of melody, but only one person in a thousand is "born with perfect pitch" sense of harmony. However, all three of these elements must be developed. Play the notes in example A. Do you recognize the melody? The reason it is difficult to recognize is because important elements: rhythm and harmony are missing.

Now play the melody with rhythm and harmony in example B. This composition has charmed music lovers for well over a century and a half, as it was published in 1801, composed by Beethoven as Sonata Opus 27, no. 2. What a great musical picture Beethoven made out of an otherwise uninteresting melody.

What is harmony? It is sometimes defined as the clothes worn by the melody. The same melody could wear many different clothes. Music is a language of emotions. It probably began with early man expressing crude emotions by the use of rhythms; drums and dances. Later shouts were added. Loud, high sounds expressed excitement while low sounds expressed grief. Thus, melody was born. Later came polyphonic music...two or more melodies at the same time, followed by the homophonic form...single line melody with a harmonic accompaniment. With the beginning of the later form, harmony was used to define the melodic tones.

A single letter of the alphabet has little meaning. Words must be formed. For example: if the letter A is used with C and E it forms the word ACE. Thus triads and chords can also be thought of as words...musical words that form our musical vocabulary. When a person's vocabulary is limited, understanding is limited. At the turn of the century even the unabridged dictionaries had less than 100,000 words. Now the latest editions contain over half a million. Imagine trying to explain the intricate workings of computers and being confined to words that were in the dictionary before 1900. This probably explains why some people do not like modern music because they are not familiar with the new vocabulary.

When should the study of harmony begin? When does the study of rhythm and melody start? Why does a student have to wait until at the conservatory level before harmony is considered? Eliminating harmony is missing one-third of musical education and two-thirds enjoyment. When a composer harmonizes a tonic note with the subdominant, the question should be "Why?" Why does the C triad sound restful in the key of C, active in the key of F and plaintive in the key of G? Music is much more exciting and interesting with the knowledge of chords and harmony.

Trying to learn chords by endless charts and non-theory combination gimmicks is tiresome and impractical. Chord application is learned quickly by practicing the common chords in a given key and then applying the same to a simple standard song or an uncomplicated melodic line. This makes the study of chords/harmony effective because results are realized quickly. The ability of a student increases dramatically as chord/harmony knowledge increases. This can be compared to an almost blind person struggling to find their way, while a person with 20/20 vision has a clear path and can see in advance. The rewards from chord/harmony knowledge is enormous; it will require patience, discipline and a knowledgeable experienced teacher to guide you.

SONATE
Sonata Quasi Una Fantasia
1801

Example A

Opus. 27 Nr. 2

TRIADS

There are four basic types of triads (Chords composed of three different letter names) (Notes)
Major, Minor, Diminished and Augmented. In popular music of the 70's a variation of one of these occurs often.
It is the suspended chord. Example one shows all five types of triads.

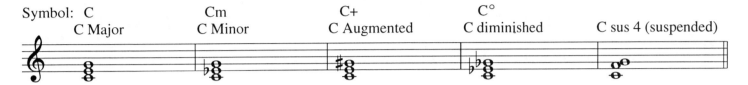

Symbol: C   Cm   C+   C°
   C Major   C Minor   C Augmented   C diminished   C sus 4 (suspended)

Chords are built from scales or intervals. Only the C major scale is necessary for this example. Remember,
this is not a theory book, it is focused on chord memorization and applications. This is all you will need to accomplish
the memorization and applications goal.

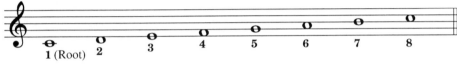

1 (Root) 2 3 4 5 6 7 8
Degrees of the Major Scale

The Major Triad = 1st, 3rd, 5th Degrees of the Major Scale

    C ( ) ← M is not needed for Major Chords.

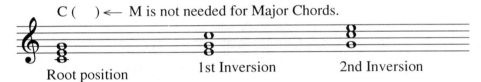

Root position   1st Inversion   2nd Inversion

Root positions are equally spaced. (Line, Line, Line) or (Space, Space, Space)

The bottom note of a chord in Root position is the letter name of the chord.

(C Minor)
   Cm = 1, ♭3, 5, of the Major Scale (Flat the 3rd)

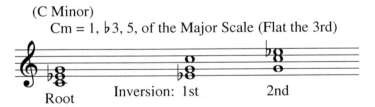

Root   Inversion: 1st   2nd

    C7 = 1, 3, 5, ♭7 of the Major Scale (Flat the 7th)

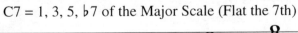
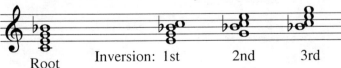

Root   Inversion: 1st   2nd   3rd   Full Title: Dominant 7th

A 3 part Chord has 3 Positions (inversions)

A 4 part Chord has 4 Positions (inversions)

Beginning lead sheet practice: Take your time. . . . Discipline & Patience.

Practice inversions only to "Get the Feel" of each chord.

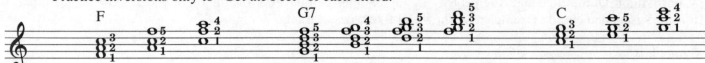

(Important: suggested fingering will be automatic later)

C = Play C Major triad notes below each written melody note.
(Do not write in the harmony notes)

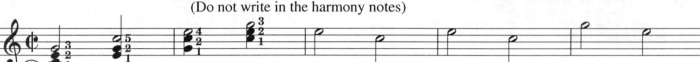

* Fingering: Always use thumb for the bottom note of the chord.

Bassoon 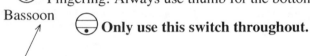 **Only use this switch throughout.**

The 1st two measures include the harmony notes in black to help you get started. The following measures require the correct harmony notes added using **memory**.

Play F Major Triad Notes (F Major Harmony)

Play G7 Chord Notes...(4 Notes)   C (Back to C Major)

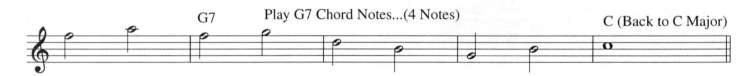

G7 chord adds "F" to the G Major triad.

F (Flatted 7th of G Major Scale)

G Major Triad

An easy method to find the 7th: The 7th of a
7th chord is <u>one whole step</u> below the octave.

octave   Remember: B to C & E to F
                  are 1/2 steps

or, one whole step below the root.
e.g. The root for G7 is G.

* Attention Accordionists:
Only use Bassoon switch throughout this book. The single reed
is more effective for ear training development. Only use bass
piano for the L.H.

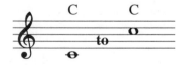

is an octave; one whole step below C is B♭
B♭ is the <u>7th</u> for C7

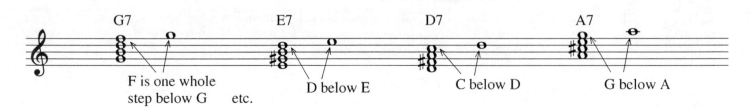

F is one whole   D below E   C below D   G below A
step below G   etc.

All exercises are mostly 8va... do not play to low; avoid muddy sounding chords.

Practice the following chords with all inversions to prepare for Excercise ①
This is not memory work; only an introduction to each chord.

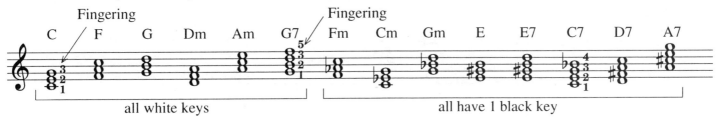

all white keys   all have 1 black key

Remember: Add the harmony below each melody note. **Do not** write in the harmony notes. ✳

(for Accordion only)

Exercise ①

Black notes are added only to help the student in beginning stages.

Remember . . . do not write in the harmony notes.

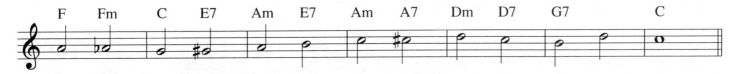

Remember to use logical fingering.   Do not add left hand until noted.

This exercise is learned when played Andante without stopping.

{ Always use the 3rd or 4th finger when
{ the Black Key is on top.

✳ Remember to read all directives and reminders

Practice first
Key of F

Practice:

Note: Play letter name of the Chord in the Bass (Root) throughout this book unless notated otherwise.

Fact: Memorization occurs only when chords are applied to an etude or melody. (Lead Sheet)

Key of Dm (Relative to F) (Am Relative to C)

Practice: Ebm Review: Gm D7 Cm

Key of Bb

Key of Gm

Practice Songs

X = Do not harmonize, only play single notes.

SONG OF THE ISLANDS

Chords and voicing to be learned later are written out

SONG OF THE VOLGA BOATMEN
(Minor Key Practice)

AUGMENTED TRIADS

The Augmented Triad raises the 5th of the Major Triad 1/2 step.
(To Augment: Raise 1/2 Step (♯))

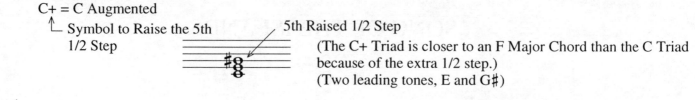

C+ = C Augmented
⌐ Symbol to Raise the 5th
1/2 Step

5th Raised 1/2 Step

(The C+ Triad is closer to an F Major Chord than the C Triad because of the extra 1/2 step.)
(Two leading tones, E and G♯)

Practice:

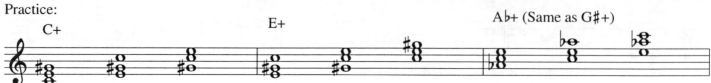

All 3 of the above Augmented Chords share the same notes!
The Bass note determines the characteristic.

same (C♯+) (Enharmonic)
     D♭+

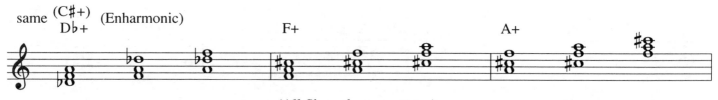

(All Share the same notes)

(All Share the same notes)

D is used instead of C✕
(double sharp)

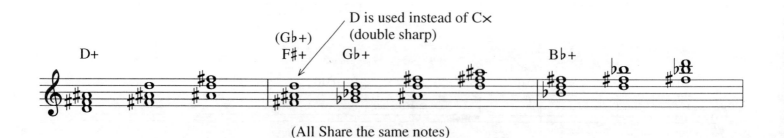

(All Share the same notes)

There are only 4 Augmented Chords!

DIMINISHED CHORDS
(Starting Below)

Diminished Chords

Diminished triad = 1, ♭3, ♭5
Diminished 7th = 1, ♭3, ♭5, ♭♭7 (Most Commonly Used)

♭♭ Never used

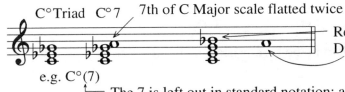

C°Triad   C°7   7th of C Major scale flatted twice

Remember C 7th is the 7th degree flatted once.
Diminished 7th means to flat the 7th of C7. (Flat B♭ once)

e.g. C°(7)
└── The 7 is left out in standard notation; always play a 4 part dim. chord for now. . . .

There are only 3 different diminished chords.

All share the same notes.
Bass note will determine chord name and or characteristic.

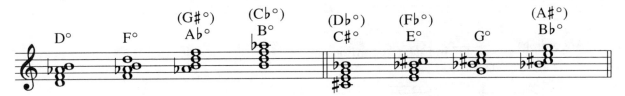

Add Harmony / Practice: (Letter name of the chord is the bottom note.) (all root positions)

& Reverse

More Practice:

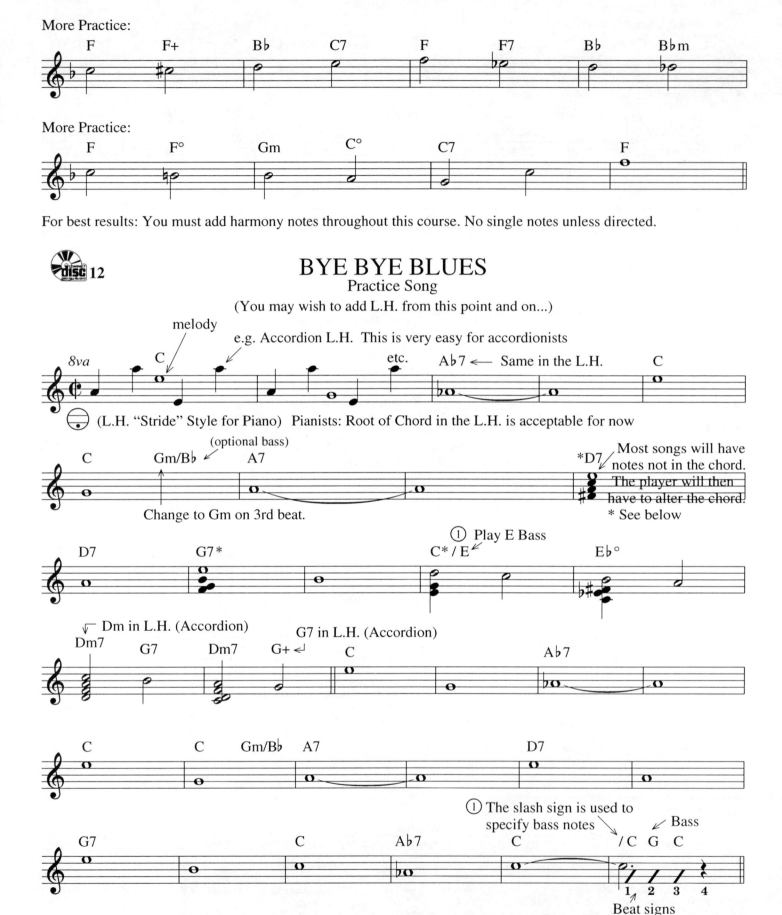

More Practice:

For best results: You must add harmony notes throughout this course. No single notes unless directed.

BYE BYE BLUES

Practice Song

(You may wish to add L.H. from this point and on...)

* Bar 9: E is not in a D7 chord. When you arrive at a non-chord note, locate the nearest position as closely as possible <u>under</u> the melody note, then simply <u>extend</u> the top note to the melody note.

LIES

Practice Songs
① See below

(Many Extensions)

Bassoon (Always use) in L.H. (Bass Piano Switch)

(Use C♯ sharp as counter bass)

MELODY OF LOVE

(C7) 1st beat only
* tacet

② Practice: Gm7

tacet
Play G Alternating Bass

Play C Bass with Gm7 Chord

① D7+5 Play a D7 Chord with the Augmented 5th.

*tacet: Play single notes only....
or solo voice.

② Practice: Gm7

DRIFTING & DREAMING

Practice Song

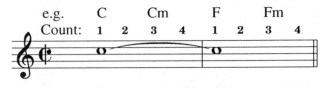

Two chord symbols in the same
measure will divide the counts equally.

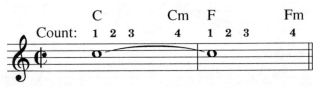

Notation with the symbol farther to
the right indicates to play the chord
on the last count.

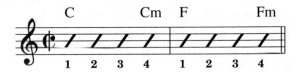

Example of chord symbols only with slashes
representing counts.

LINGER AWHILE

Practice Songs

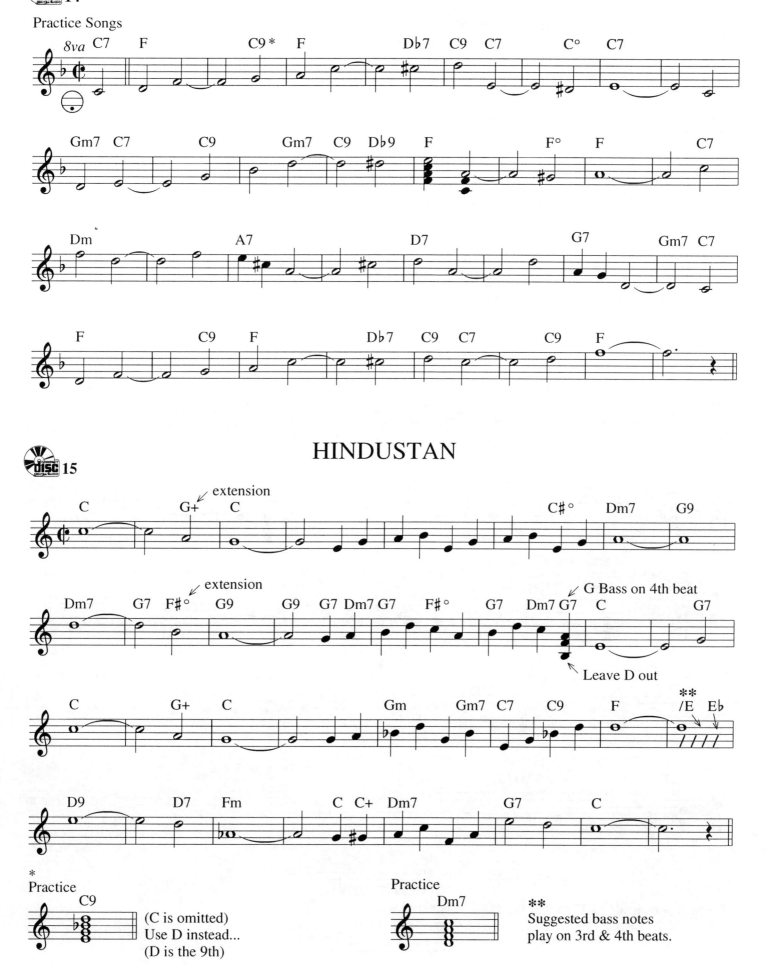

HINDUSTAN

THE MAJOR 6TH CHORD

The Major 6th Chord = 1, 3, 5, 6 (4 part chord)
The 6th is added to the triad for a thicker sound (Big Band style).

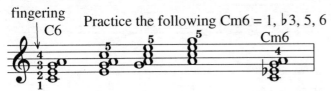

Practice the following Cm6 = 1, ♭3, 5, 6

Practice all new chords in inversions

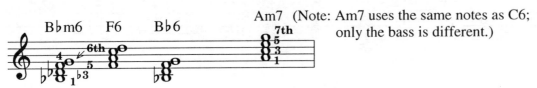

Am7 (Note: Am7 uses the same notes as C6;
only the bass is different.)

Also Practice:

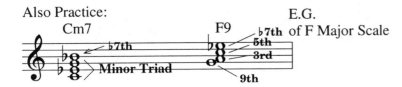

E.G.

♭7th of F Major Scale

BYE BYE BLACKBIRD

Play F Bass with B♭6

A Bass with F6

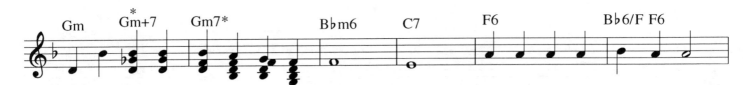

Note: Cm6 and F9 share the same notes.

The bass note, again, defines the chord.

* Explained later in the course.

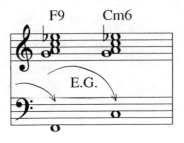

① Play A Bass on
1st & 3rd count

② Play A♭ Bass on
1st & 3rd count

16

THE MINOR 7TH CHORD

E.G. Play Cm in L.H. (not C7) (Accordion)

The Minor 7th Chord = 1, ♭3, 5, ♭7
A Minor triad with a little
7th chord flavoring.

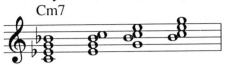

Practice the following m7 chords using all inversions:

Cm7   Gm7   Dm7   Am7   Fm7   Em7   B♭m7   E♭m7   A♭m7

E.G. The Cm7 chord is a preliminary and passing chord for the
7th chord a 5th below.

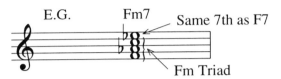

Same 7th as F7

Fm Triad

Remember. . . play roots in the bass

Practice: (all 4 part chords)

Note the 6th interval on the bottom note to avoid dissonance using m7 chords.

The same melody
using only G7. . .
note the difference.

(Listen) (Dissonant?)

Now play the same melody
using 6th intervals
    (2 notes only) →

No Dissonance

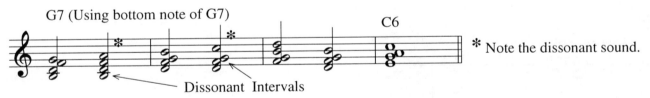

G7 (Using bottom note of G7)

Dissonant Intervals

* Note the dissonant sound.

Practice:

Note: The m7 passing chord used for each 7th chord is always a 5th higher.
    (accordion basses are organized in 5ths)

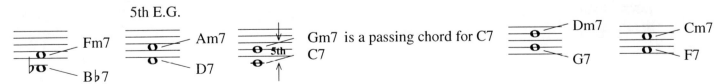

5th E.G.

Fm7   B♭7   Am7   D7   5th   Gm7 is a passing chord for C7   C7   Dm7   G7   Cm7   F7

17

PEG O' MY HEART
(This song is usually harmonized poorly)

(Use all 4 part chords unless noted throughout the course)
Easy going speed. Bass clef is added to demonstrate L.H. usage.

Note how ╱ (slash) signs match up with L.H. bass notes.

PRACTICE ALL CHORDS WITH INVERSIONS

Chord Review:

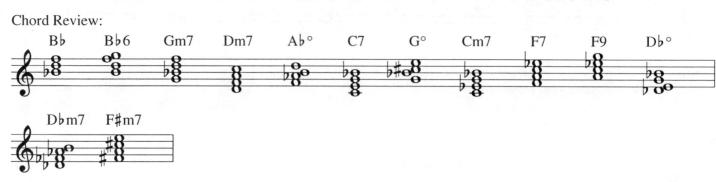

SWANEE

Remember: Add harmony for all single notes.

① All 4 part chords except + chords and no root chords.

② F+ and A+ use the same notes: Using A in the bass changes the characteristic of the chord.

19

THE MAJOR 7TH CHORD

The Major 7th Chord = 1, 3, 5, 7

Notation: C maj(7) used on some lead sheets

Some lead sheets will notate as: M7 or △

7 not necessary but

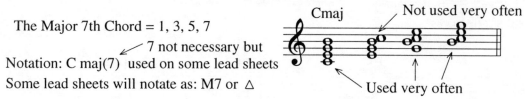

For a less dissonant sound, the root is often left out from the treble and played in the bass using octave voicing or 3 part voicing.

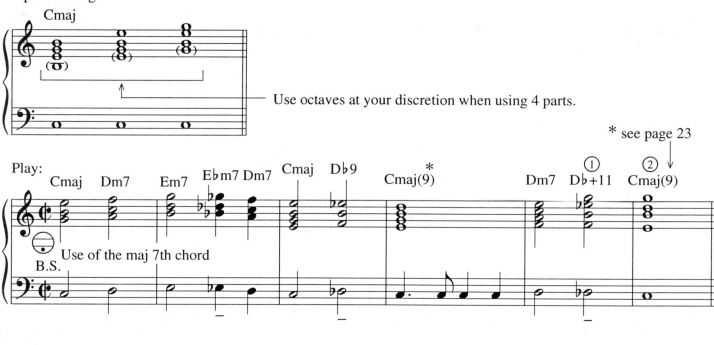

Use octaves at your discretion when using 4 parts.

* see page 23

Play roots in the bass & omit in the treble.

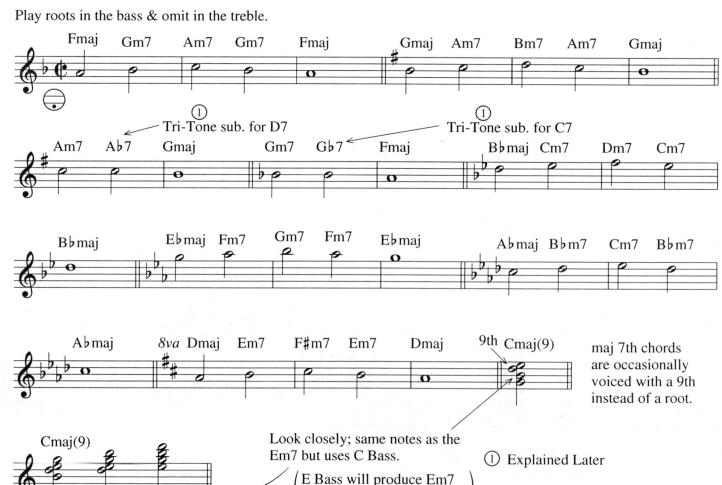

maj 7th chords are occasionally voiced with a 9th instead of a root.

Look closely; same notes as the Em7 but uses C Bass.

(E Bass will produce Em7
  C Bass will produce Cmaj9)

① Explained Later

THE DOMINANT 9TH CHORD

The 9th chord is usually played without the Root.

C9 = 1, 3, 5, ♭7, 9
Symbol: C9

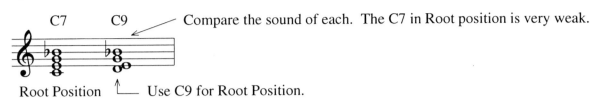

C7   C9   — Compare the sound of each. The C7 in Root position is very weak.

Root Position   ⌐— Use C9 for Root Position.

Only use the 7th in 1st and 2nd Inversions.   Do not use the 9th in 1st and 2nd inversions.

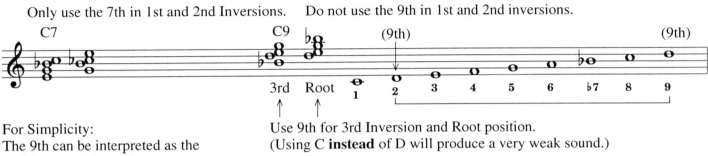

For Simplicity:
The 9th can be interpreted as the
2nd degree of the scale.
(One whole step above the Root.)

Use 9th for 3rd Inversion and Root position.
(Using C **instead** of D will produce a very weak sound.)

Practice the following 9th chords with all inversions ① only for practice.

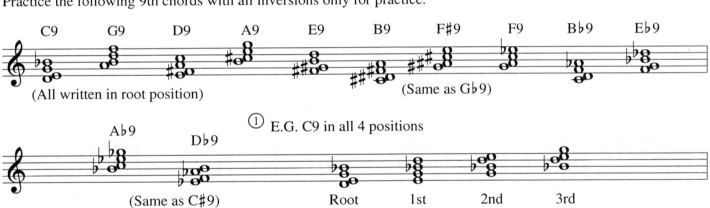

C9   G9   D9   A9   E9   B9   F♯9   F9   B♭9   E♭9

(All written in root position)   (Same as G♭9)

A♭9   ① E.G. C9 in all 4 positions
        D♭9

(Same as C♯9)   Root   1st   2nd   3rd

(+ or ♯)   (- or ♭)

RAISED AND FLATTED NINTHS

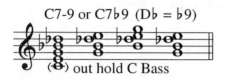

C7-9 or C7♭9 (D♭ = ♭9)

(⊕) out hold C Bass

Find and practice inversions for ♭9th chords from
the 9th chord study on the previous page.

Note: The notes of a C7-9 are exactly the same as G°

Remember: Bass note determines characteristic of the chord.

Hold root basses on all -9 & +9 practice.

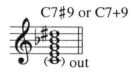

C7♯9 or C7+9

(⊕) out

This chord is used in root position most often for a cleaner sound.
Find and practice all +9th chords from the 9th
chord study on the previous page.
Use the 3rd of the chord as the bottom note... Root position only.

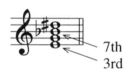

7th
3rd

The 3rd and ♭7th is essential for all
7th, 9th, ♭9th, ♯9th, 13th dominant chords.
    (-9)   (+9)

| 11th and +11 dominant chords usually omit the 3rd. |

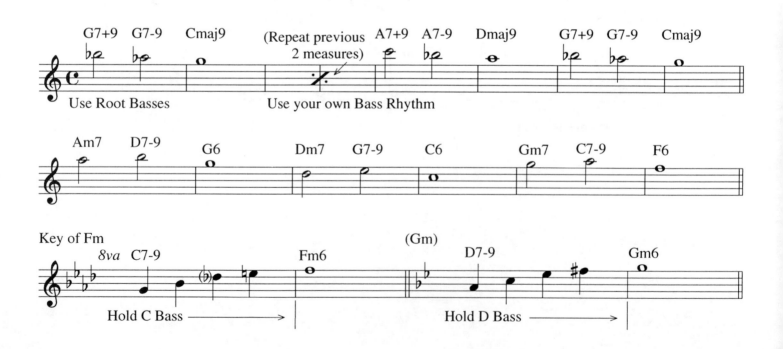

DOMINANT 11th AND AUGMENTED 11th CHORDS

+11

C11

Formation

Symbolized as C11

9th, 7th, 5th while the 3rd and Root are often left out.

Also Contains:

C Scale in 2 Octaves

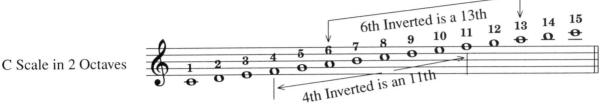

6th Inverted is a 13th

4th Inverted is an 11th

As played   C11   (Gm7/C)

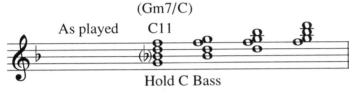

Hold C Bass

The C11 chord may be written as Gm7/C Bass, using m7 chords and a 5th below bass simplifies memorization.

Memorize circle of 5ths

(F♯)   (G♭)   → (D♭) (A♭) (E♭) ← (enharmonic)

G♭ D♭ A♭ E♭ B♭ F C G D A E B F♯ to Continue: C♯   G♯   D♯ etc.

It is called a circle because it meets at G♭/F♯

B♭♭ F♭ C♭ ← G♭   Note how the pattern repeats the letter name.

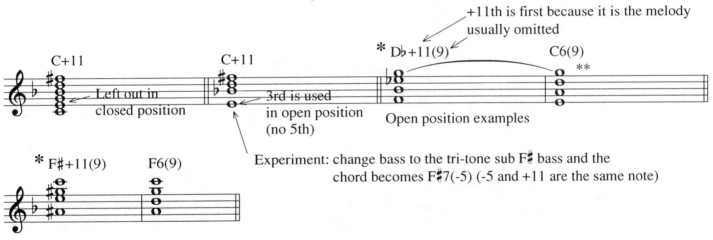

+11th is first because it is the melody usually omitted

* Db+11(9)   C6(9)

C+11 — Left out in closed position

C+11   3rd is used in open position (no 5th)

Open position examples

**

* F♯+11(9)   F6(9)

Experiment: change bass to the tri-tone sub F♯ bass and the chord becomes F♯7(-5) (-5 and +11 are the same note)

Eb7 with Eb bass and becomes A7 with A bass.

* Example of a Tri-tone substitute

A Tri-tone is 3 Whole steps.
Mainly used with dominant chords.
The Tri-tone creates a 1/2 step resolution.

G7+5   Db+11(9)   C6(9)

The Tri-tone movement is used often in jazz.
Note the interesting "Color" added to the harmonic progression.
G7 to C is more interesting as G7, Db7, C.

Same notes, only the bass changes on the Tri-tone.

Note the Reversal: B & F of G7 become F & B of Db7!
(Same notes)   3rd   7th   3rd   7th

** Occasionally a ( ) will be added to the chord symbol for note clarification if a particular chord is preferred. (or as a courtesy)

OPEN POSITION CHORDS

Open Chords: Use the first note under the melody note from closed position and play as the lowest note for open position.

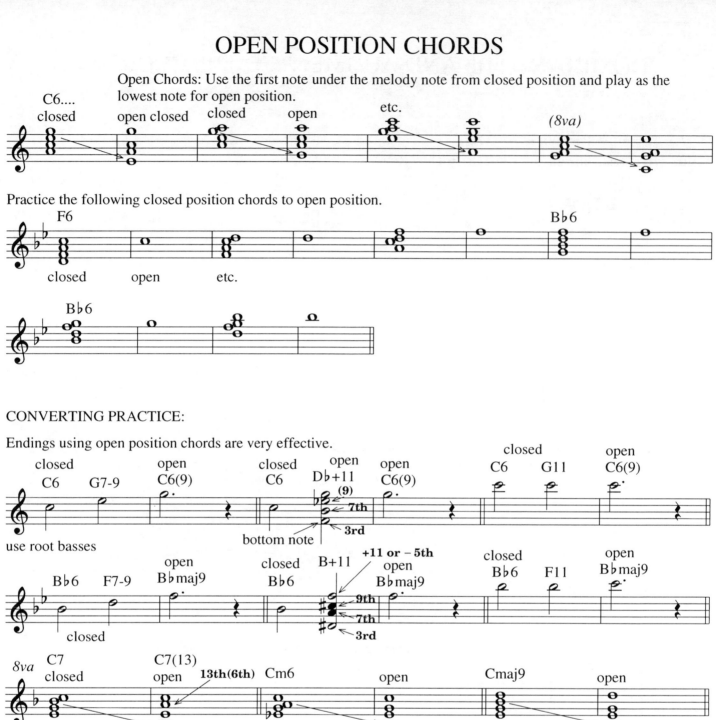

Practice the following closed position chords to open position.

CONVERTING PRACTICE:

Endings using open position chords are very effective.

use all open position:

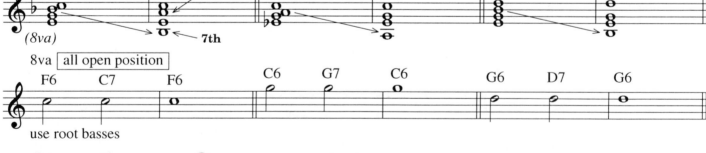

① Usually written as 13 only. e.g., C7(13) for clarification during study to denote type of chord
e.g., C13

PASSING CHORDS

Harmonizing passing melody notes: (Notes that are foreign to the chord are known as accidentals) Use the diminished chord of the same name of the chord following the accidental.

E.G.   Chord symbols in ⟶ (   ) ⟵ only for that note... Return to the primary chord following the (   ).

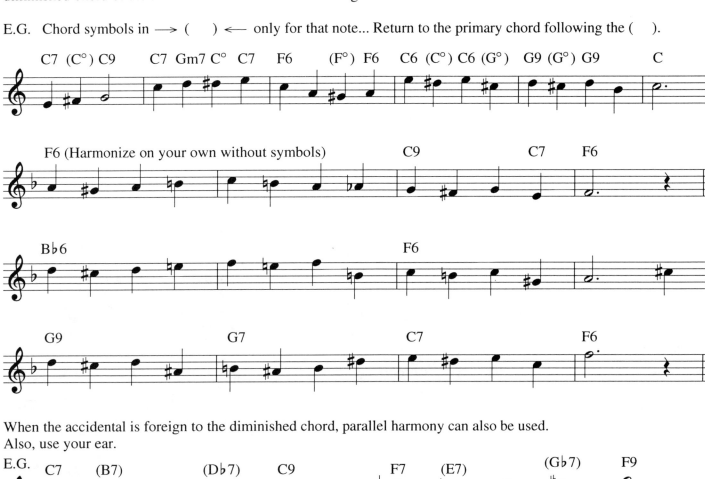

When the accidental is foreign to the diminished chord, parallel harmony can also be used.
Also, use your ear.

E.G.

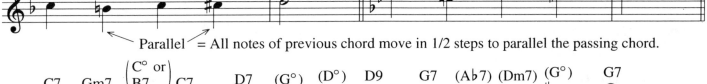

⟵ Parallel ⟶ = All notes of previous chord move in 1/2 steps to parallel the passing chord.

Harmonize using dim. chords.

*Note: Utilize a 6th interval below when planning the best sounding passing chord.
   *For 7th chords: When applicable, use the m7 chord as a passing chord for the 7th chord in the boxes.

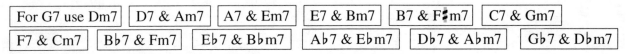

For G7 use Dm7	D7 & Am7	A7 & Em7	E7 & Bm7	B7 & F♯m7	C7 & Gm7
F7 & Cm7	B♭7 & Fm7	E♭7 & B♭m7	A♭7 & E♭m7	D♭7 & A♭m7	G♭7 & D♭m7

   * Ask your teacher to clarify.

YOU MADE ME LOVE YOU

APRIL SHOWERS
(A Few Helping Notes Written In)

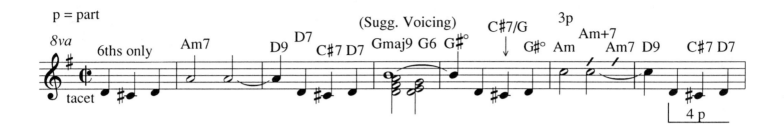

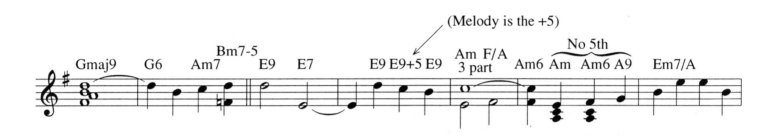

WHAT A FRIEND

Traditional

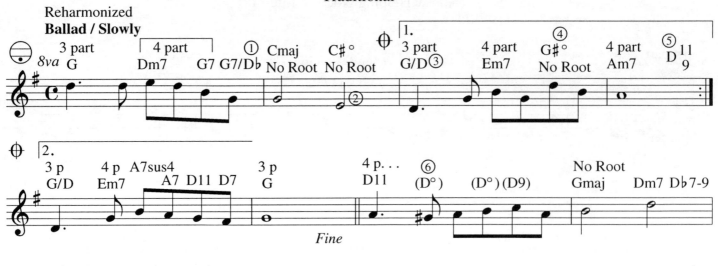

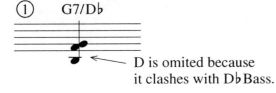

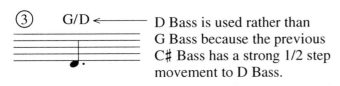

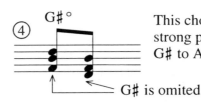

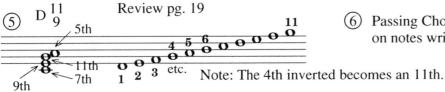

Voicing Examples

① G7/D♭ — D is omitted because it clashes with D♭ Bass.

② C♯ — C♯ is omitted for a cleaner, wide open sound and is played in the bass. This is perceived open because the bass keyboard is 2 octaves lower than treble. (Accordion)

③ G/D — D Bass is used rather than G Bass because the previous C♯ Bass has a strong 1/2 step movement to D Bass.

④ G♯° — This chord is used for a strong pull to Am7. G♯ to A is a 1/2 step. G♯ is omitted

⑤ D 11 9 — Review pg. 19 — 5th, 11th, 7th, 9th, 4 5 6 11, 1 2 3 etc. Note: The 4th inverted becomes an 11th.

⑥ Passing Chords; only played on notes written with (   ).

⑦ Another voicing for D11 — 4th is considered suspended. It is a restless note and must resolve to the 3rd(F♯) of the chord.

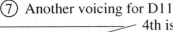

⑧ D7 — Resolved to F♯ from the Previous sus 4th (G). (G to F♯ = 1/2 Step)

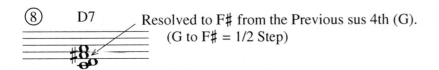

STUMBLING

① This piece is a good example of many single note passages because of the bright tempo and eighth note figures.

② Many lead sheets will only state the basic harmony. The B♭ melody actually produces a D9+5.

③ Use F as bottom note. The actual chord is C7sus; F is the suspended 4th, (or, it could be a Gm7/C).

④ Basic C7 best here ⑤ Use G9 ⑥ Basic G7 ⑦ Use C9 ⑧ Just happens to be the same notes as C6.
F bass produces Fmaj9.

⑨ This ending compliments the song's style and extra harmony has been added to the original to produce more color.

SWEETHEART OF SIGMA CHI

PRACTICE SONG

3 Part Chords

* Use 3 part chords on Major & Minor & some 7th Chords.

** Find best passing chord; bottom is written to help out.

① Which interval sounds better? This is why G6 for G7 harmony is used on the E melody note when using 3 part chords.

FASCINATION

Example of a "Turn Around", melody line & chord progression same as an introduction.

① Learn to "Feel" a slide from chord to chord.

② Only the top & bottom notes rotate down — Hold the middle triad.

③ 5 Part chords are mainly played as octaves (Known as a double lead; i.e. the melody is on top and also on the bottom) big bands use this effect, but use a few more octaves. This produces a very big/rich sound and at times more emotional sounding chords.

④ This is easy – Only 1 notes changes; the C of Dm7 moves to B of G9.

⑤ 3rds sound appropiate here.

⑥ Pros often play:

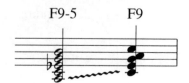

CHINATOWN

① Bass pattern often used with a long series of the same harmony.

② The major 7th for Λ minor chord is usually symbolized Am+7. (Am maj7 is not used) (Fm maj7 is not used.)
+7 is used for the maj7th note on minor chords only.

JA - DA

Practice song with a standard intro. Your choice on 3 or 4 part chords....
Use a variety.

(Remember to use passing chords)

for Repeat D.S.

End only (Standard ending to fit the Tune)

ala Welk

No Bass

Play all   as   E.G.

Count: **1 & a 2 & a 3 & a 4 & a**

In Pop, Jazz, Dixie, etc. eighth notes are usually played with a more relaxed rhythm.

Not like this:
Count: **1e& a 2e& a 3e& a 4e& a**

This note is too quick before
the next beat for Jazz & Pop.

AFTER YOU'VE GONE

A good modulation example
for a key change

A minor 7th with an
added 11th because of
the melody (no 5th).

Suspended 4th
D will resolve to the 3rd.

③ F♯m7 passing chord

This song starts in F and ends in C

DO IT AGAIN

(Early Gershwin)

(Practice Tune)

① Good example of internal movement. Note the movement.

* ② Ab7 is a Tri-Tone sub. for D7

How we find the Tri-tone subs.

Db is the flatted 5th of Ab7-5

Ab is the b5th of D7-5

* ② Review 3rd & 7th of Ab7 and D7, both chords share the same 3rd & 7th in reverse order, also known as the flatted 5th sub. (Tri-tone sub.)

MY MELANCHOLY BABY

Ad lib your own ending melody line using the chord progression.

* Note the effective use of the flatted 5th.

I'M JUST WILD ABOUT HARRY

① It would be rather redundant to only stay with Am for 2 measures.

    The +7 and m7 add
      (G♯) (G♮)
    Movement and Color

INDIAN SUMMER

3 Methods to harmonize this section.... Choose the one you prefer.

Advanced left hand combinations

②Gmaj = G bass & DM Chord ③ A♭9+11 = G♯ bass & D7 ④ Am7 = A bass & CM
⑤F♯m7 = F♯ bass & AM ⑥Am7-5 = A bass & Cm

CHORD STUDY AND REVIEW

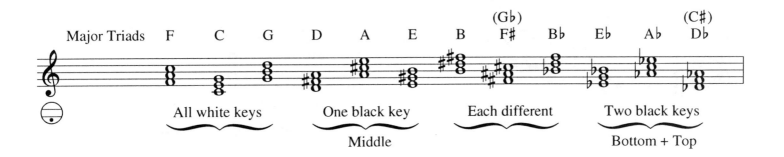

| Major Triads | F | C | G | D | A | E | B | (Gb) F♯ | Bb | Eb | Ab | (C♯) Db |

All white keys — One black key (Middle) — Each different — Two black keys (Bottom + Top)

Varieties of the triad using C and Cm as the example.

8va C-5 Csus4 Csus2 C+ Cm Cm+5 Cm6 C6 Cm6 C6 C° Cm9 Cm+7 Cm9

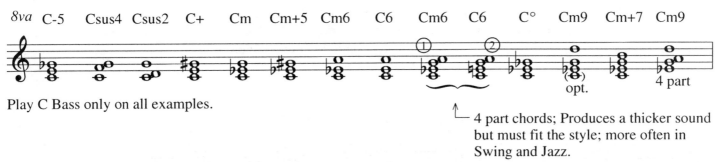

(8) opt.   4 part

Play C Bass only on all examples.

4 part chords; Produces a thicker sound but must fit the style; more often in Swing and Jazz.

Varieties of the dominant chord using C7 as the example:

(Half diminished 7th)

7th is always omitted for diminished.

3rd of the chord usually left out

C7 C9 Cm7 C⌀7 (Cm7-5) C° 7th C7-9 C7+9 C11(9) C+11(9) C13(9)

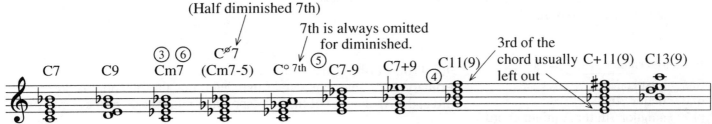

(continued on pg. 40)

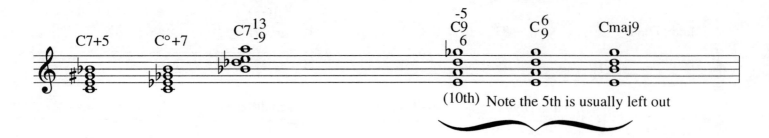

(10th) Note the 5th is usually left out

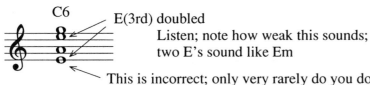

C6

E(3rd) doubled

Listen; note how weak this sounds;
two E's sound like Em

This is incorrect; only very rarely do you double the 3rd of the chord;
this also pertains to the 3rd in the bass. i.e., <u>do not</u> play E (3rd) in the bass unless
omitted in the treble.

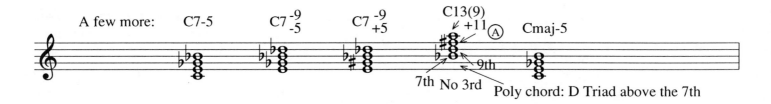

7th No 3rd   Poly chord: D Triad above the 7th

As you can see, all chords develop from Major and Minor triads.
Make sure they are memorized.

Remember: | + = ♯ || − = ♭ |

Sus = Suspended 4
maj = major 7th
(+7 is a major 7th for A minor chord.)
① Similarties: Cm6 is F9 with F Bass
② C6 is Am7 with A Bass
③ Cm7 is E♭6 with E♭ Bass

9th, 11th, +11th, 13th chords are dominant chords.
Primary color notes: i.e., Notes that determine the 7th chord is the
3rd and ♭7th degrees of the corresponding major scale of the same name.

④ Same as Gm7/C.
⑤ C7-9 is E°, G°, B♭°, D♭° with Root Basses
⑥ Cm7 is the same as $F^{11}_{(9)}$

I'M FOREVER BLOWING BUBBLES

(Bottom notes)

① Melody is on sus4   ② Tri-Tone sub for B♭7   *Remember: Use thoughtful passing chords

GEORGIA

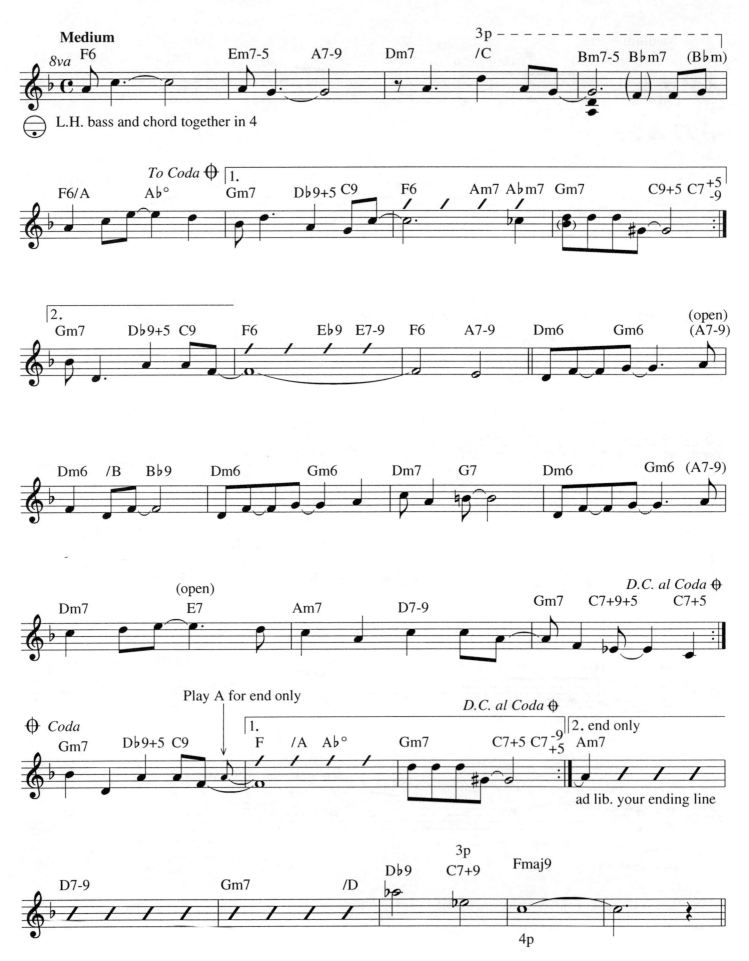

A GOOD MAN IS HARD TO FIND

① Tacet: Single note solo or voice solo.
② N.C. = No chord but bass line or L.H. chords can be added.

'WAY DOWN YONDER IN NEW ORLEANS
(12 BAR BLUES)
(PASSING CHORD PRACTICE)

3p = 3 part
4p = 4 part
Medium Blues

Written in 1922. This song is a favorite Dixie Tune.

THE JAPANESE SANDMAN

① Fill exmples

BILL BAILEY

A few pages of lead sheets for arranging skills practice.
Hint: Don't try to add too many extra chords. Use thoughtful passing chords.

* Single notes

MA
(He's Making Eyes at Me)

Two practice songs

Medium (many passing chords left out)

MARGIE

Only the primary harmony is on this chart. The student must now add passing chords
and any substitutions.

Pick up (start) for Margie

Ask your teacher how to add more rhythm.

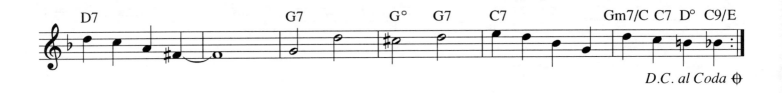

D.C. al Coda

AVALON

Add passing chords and substitutions.

AVALON WITH EXTRA HARMONY

Also see Avalon in Jazz Accordion Solos book. MB96309BCD

(More interesting version)

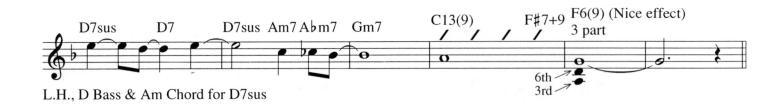

L.H., D Bass & Am Chord for D7sus

ALEXANDER'S RAGTIME BAND

INDIANA

INDIANA
(Reharmonized)

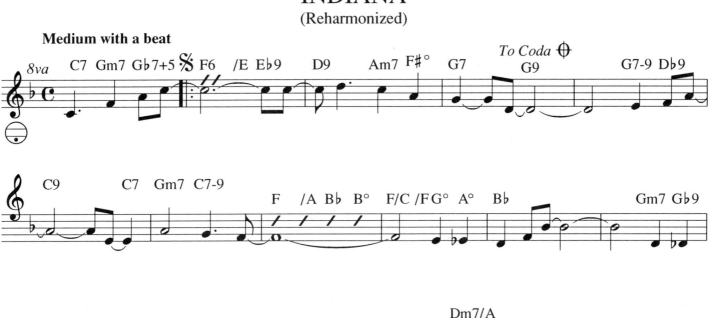

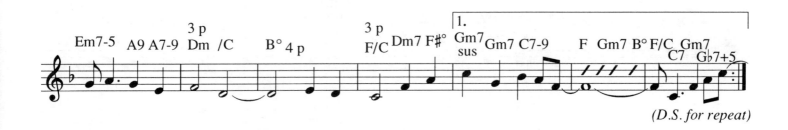

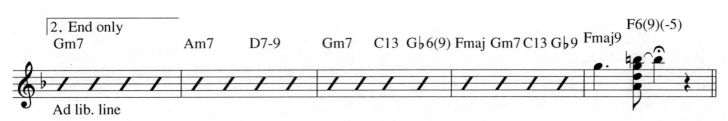

Also see "Indiana" in the book *Jazz Accordion Solos*. MB96309BCD

POOR BUTTERFLY

Passing Chords, Chord Fills, Complex Harmony

Also see "Poor Butterfly" in *Jazz Accordion Solos*. MB96309BCD

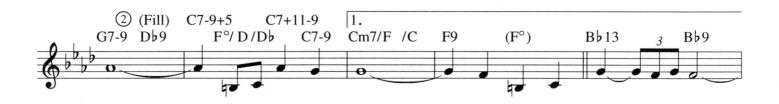

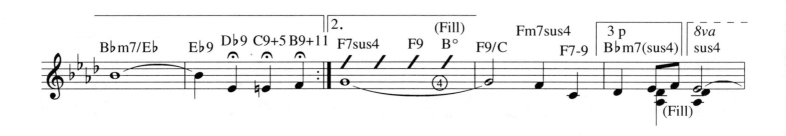

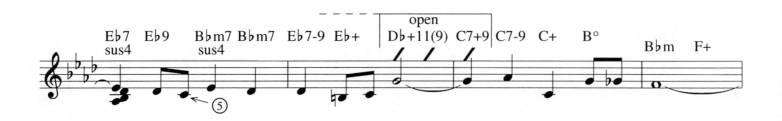

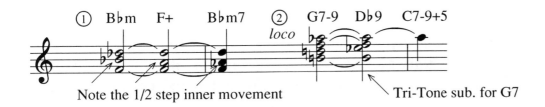

Note the 1/2 step inner movement

Tri-Tone sub. for G7

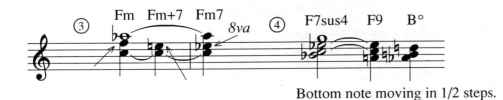

Bottom note moving in 1/2 steps.

F is bottom note for both melody notes.

Important for more practice songs

For more practice songs purchase a published fake book with correct harmony. 'Fake Book' is a slang title for a book only with lead sheets ...... same as in this course.

About the Author

Gary Dahl is widely known as a virtuoso accordionist as well as a composer, arranger, recording artist and music educator, with an extensive background in music theory, composition and harmony. Gary has now developed an impressive body of work including hundreds of individual arrangements and more than a dozen books currently in print with Mel Bay Publications.

As a recognized teacher, Gary provides specialized training for all levels of students. Gary's students have won national and state competitions as well as achieving professional status. While Gary resides in Puyallup, Washington near Seattle, he provides lessons by correspondence for students worldwide. **www.accordions.com/garydahl**

Gary currently performs as a single for private functions. The Gary Dahl trio plus vocalist performed regularly at private clubs, hotels and the lounge circuit from 1960 through 1991. Gary is a graduate of the University of Washington specializing in composition and theory and is a former commercial corporate pilot, flight instructor and corporate sales manager.